God's Cool Creation:

Epic Earth

GOD'S COOL CREATION: EPIC EARTH (BOOK SERIES, BOOK 1)

First paperback edition: June 2022

ISBN: 979-8-218-01432-2
LCCN: 2022910330

Written and illustrated by Mary Ann Winslow
Contribution by Benjamin Winslow

Disclaimer: This book contains general science information that is intended to help the reader better understand basic principles. It is understood that some specific details have been omitted. For example, the exact mileage from Earth to the moon is rounded off, the minor contributions of the Sun and other factors when discussing tides are not included, etc.

This book is dedicated to Henry
and his grandpa

Earth's tilt is important, too!
It's a perfect 23.5 degrees.
It gives us our....

Bragger!!

I'm amazing!

I am, too!

No, you're not!!

SEASONS!

No tilt? It would be really cold in a lot of places, and....

Earth's moon is important, too!

Hey! You forgot me!

OOPS! Sorry!

It's 240,000 miles away. According to www.spaceplace.nasa.gov, that's 32 Earths lined up! But that's just right because it keeps Earth from wobbling! And here's a question for you...

I'm gunna be sick!

Did you ever wonder why we always see the Man in the Moon? It's because it takes the same time for the moon to spin around (rotate) as...

I'm spinning!

WOW!

it takes the moon to go around Earth (revolution)! So Earth's month is the same as the moon's day!

The moon's size matters, too. It's big! It's about one fourth the size of Earth!

God timed all this
perfectly for us!
Because if we...

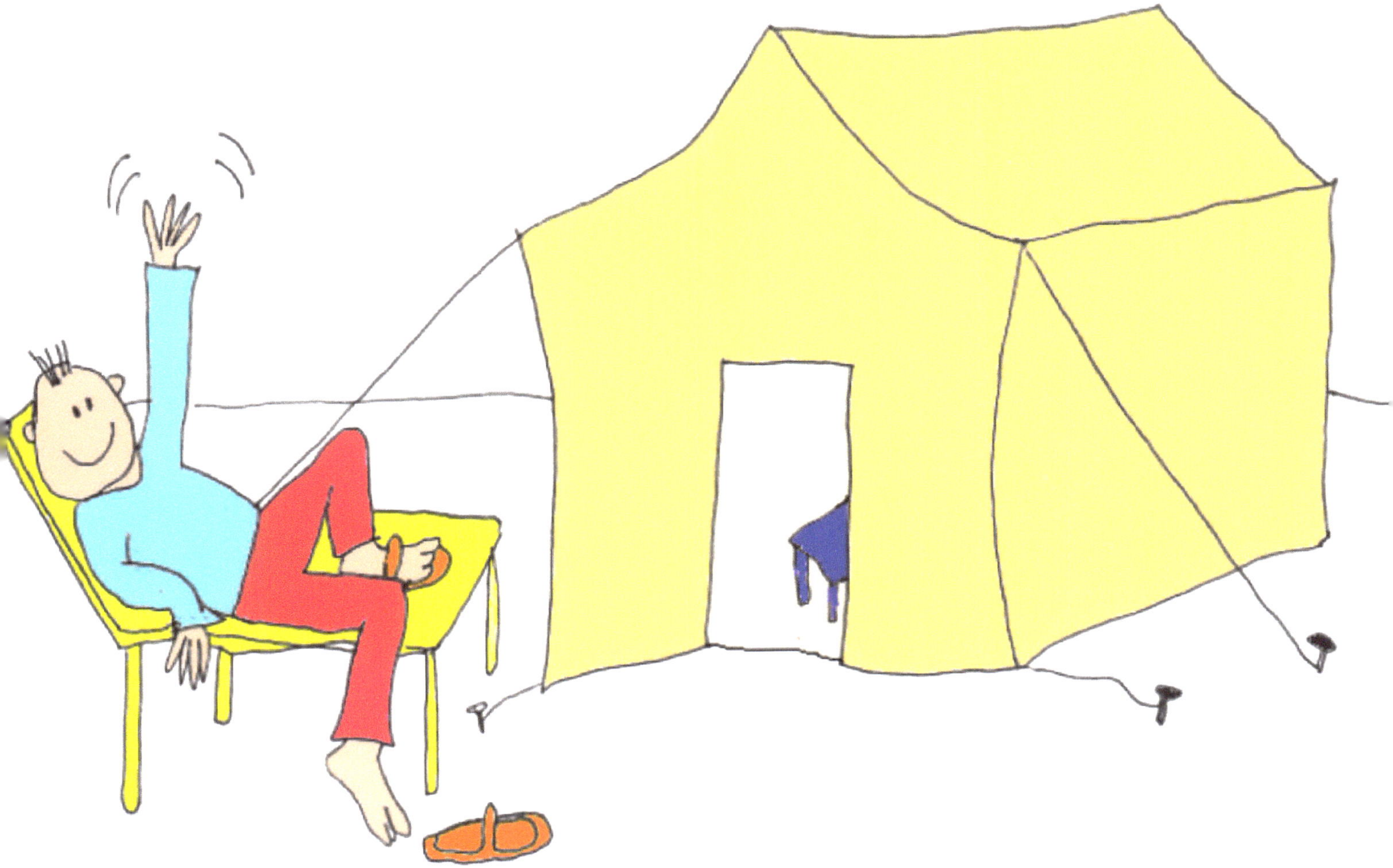

...add up the moon's size, spin, distance from Earth, and the pull on Earth, we get...

Well, the moon pulls on Earth just enough to make our oceans move!

I'm sailing!

The sea rushes to shore and falls back again to the sea (mostly twice a day).

Sea creatures love this because tides bring food when they dash to shore!

They also send cold water to really hot places and...

Ahh! This is the life!

...warm water to really cold places, making Earth livable!

Tides even help sea turtles travel! Momma sea turtles fling to shore to lay eggs, thanks to tides.

And they hurl baby sea
turtles back to the sea!

God works ALL this perfectly
just for us!

Thank you, Lord!

Watch for more God's Cool Creation books coming soon!

God's Cool Creation: Wonderful Water
God's Cool Creation: Vivid Volcanoes
God's Cool Creation: Dazzling Desert
God's Cool Creation: Terrific Trees
God's Cool Creation: Amazing Arctic
God's Cool Creation: Sensational Sea Mammals
God's Cool Creation: Super Cells
God's Cool Creation: Lovable Leaves

About the author: Mary Ann Winslow, PhD, is a former science teacher, university instructor (University of Wyoming, Texas A&M University, University of Arizona), SIU Saluki, UofA Wildcat (Bear Down!) and most importantly, Jesus follower. She currently resides in Prescott, Arizona, with her husband Kent.

www.ingramcontent.com/pod-product-compliance
Lightning Source LLC
Chambersburg PA
CBHW041441120626
46547CB00002B/291